Examining
Oil and Coal

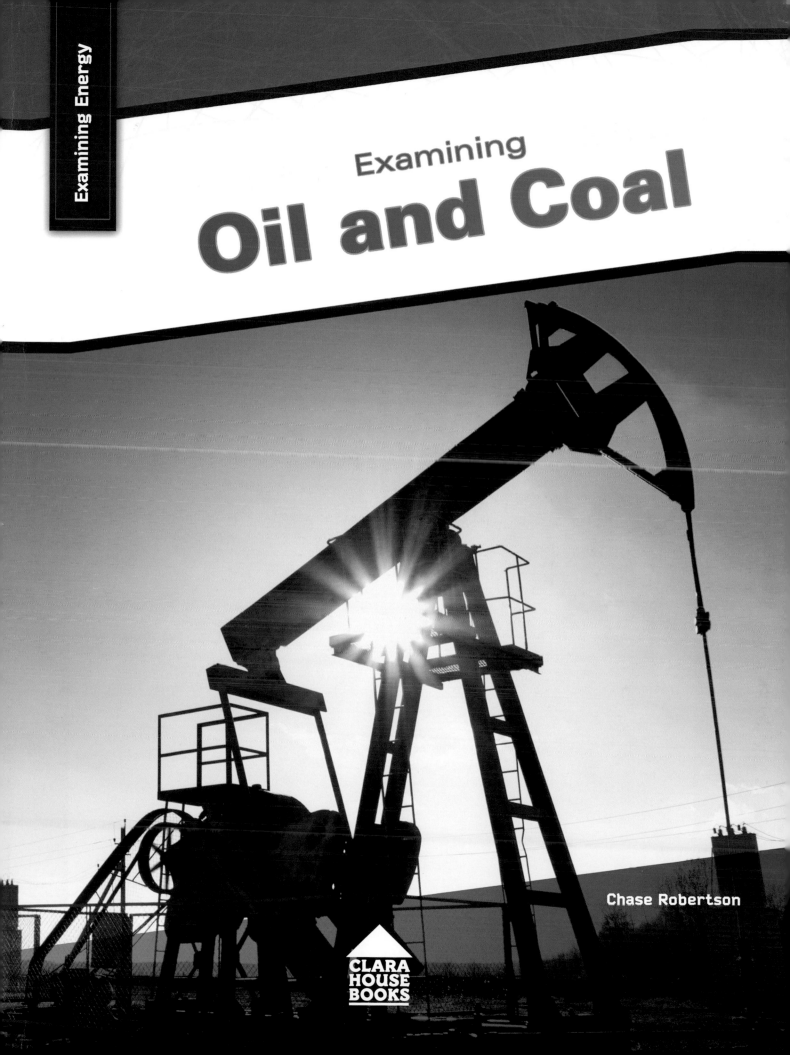

Chase Robertson

CLARA HOUSE BOOKS

First published in 2013 by Clara House Books, an imprint of
The Oliver Press, Inc.

Copyright © 2013 CBM LLC

Clara House Books
5707 West 36th Street
Minneapolis, MN 55416
USA

Produced by Red Line Editorial

The publisher would like to thank Burtron H. Davis, Associate Director, Clean
Fuels and Chemicals, Center for Applied Energy Research, University of Kentucky,
for serving as a content consultant for this book.

Picture Credits

Fotolia, cover, 1, 6, 11, 25, 41; Jaydee Jana/iStockphoto, 5; AP Images, 8; Cheryl
Casey/Shutterstock Images, 13; Bigstock, 15; Tomas Sereda/Fotolia, 16-17; Juan
Camilo Bernal/Shutterstock Images, 19; Shutterstock Images, 21, 38; Red Line
Editorial, 22; Przemek Tokar/Shutterstock Images, 23; James Phelps/Bigstock, 27;
Library of Congress, 29; Vitoriano Jr./Shutterstock Images, 30; Mark Winfrey/
Shutterstock Images, 32; Nathalie Speliers Ufermann/Shutterstock Images, 33;
Peter Close/Shutterstock Images, 36; Steve Jacobs/iStockphoto, 39; Catherine
Yeulet/iStockphoto, 45

Every attempt has been made to clear copyright. Should there be any inadvertent
omission, please apply to the publisher for rectification.

Library of Congress Cataloging-in-Publication Data

Robertson, Chase.
 Examining oil and coal / Chase Robertson.
    pages cm. -- (Examining energy)
 Audience: Grades 7 to 8.
 Includes bibliographical references and index.
 ISBN 978-1-934545-44-7 (alk. paper)
1. Petroleum as fuel--Juvenile literature. 2. Coal--Juvenile literature. I. Title.

 TP355.R63 2013
 553.2'82--dc23

                          2012035249

Printed in the United States of America
CGI012013

www.oliverpress.com

Contents

Chapter 1	**Energy for the Future**	**4**
Chapter 2	**Oil Wells**	**7**
Chapter 3	**Offshore Drilling**	**10**
Chapter 4	**Turning Oil into Energy**	**14**
Chapter 5	**Into the Seam**	**20**
Chapter 6	**Coal Energy**	**26**
Chapter 7	**Cleaner Coal**	**31**
Chapter 8	**A Bright Future?**	**35**
Chapter 9	**Your Turn**	**40**

Glossary	**42**
Explore Further	**44**
Selected Bibliography	**46**
Further Information	**47**
Index	**48**

CHAPTER 1

Energy for the Future

Have your parents complained about the high cost of gasoline? What about the high cost of electricity? People all over the world use a lot of energy. Right now, a lot of the world's energy comes from non-renewable sources. Obtaining and using these non-renewable sources, such as oil and coal, can have negative effects on the environment. And they will eventually run out. Energy keeps getting more expensive as world populations and economies grow and supplies shrink.

Scientists are constantly looking for ways to improve our sources of energy. Right now, fossil fuels including oil and coal make up more than 80 percent of the U.S. energy needs. They are likely to continue being the sources for the majority of our energy for many years to come. Finding ways to make oil and coal more environmentally friendly, less expensive, and last longer are important parts of meeting our energy needs of the future.

Fossil fuels are the remains of plants and animals that died millions of years ago. These fuels take a very long time to form. This is what makes them non-renewable energy sources. Fossil fuels include coal, oil, and natural gas. Coal is a black or brown rock made up mostly of carbon. Burning it as fuel releases large amounts of energy. Most coal in the United States is used to generate electricity.

Crude oil, or petroleum, is a liquid fossil fuel. It is made of carbon and hydrogen. Petroleum is converted to gasoline, diesel, and jet fuel, which are our primary sources of energy for transportation. Finding ways to mine and

Gasoline made from oil is a very important part of our modern way of life.

use coal and oil more efficiently will play a major role in our energy future.

Oil comes from ancient creatures turned into fossil fuels after millions of years of heat and pressure.

EXPLORING FOSSIL FUELS

In this book, your job is to learn about how coal and oil are used as energy sources and their role in our energy future. How were these energy sources formed? How are we using them? How do they produce energy? What can we do to make them more efficient and clean?

Jamaal Carter is traveling to key destinations as part of a field team studying coal and oil. He will meet with experts in the field to help him learn more about oil and coal and how these energy sources fit into our future. Reading his journal will help you with your own research.

CHAPTER 2

Oil Wells

My team and I kick off our research by learning about oil. We've come to an oil well in Texas. Adam Jacobs, the vice president of the oil company, meets us when we arrive. First he tells us a little bit about the oil his company is drilling for. He explains that oil is a fossil fuel. It formed from tiny sea creatures that lived more than 300 million years ago. When these sea creatures died, they sank to the bottom of ancient oceans and became covered with mud. The layers became thicker until there were hundreds and sometimes thousands of feet of earth covering the dead sea life. Over millions of years, pressure and heat turned the sea life fossils into oil and natural gas.

Mr. Jacobs tells us that Edwin L. Drake in Titusville, Pennsylvania, drilled the first oil well in 1859. At first, oil was used for kerosene in lamps. Later, by the end of the 1880s, more and more cars were produced, and they ran on gasoline. By the 1920s, there were nine million cars in the United States. Gas

Edwin L. Drake, *right*, drilled the first oil well in 1859.

stations began popping up everywhere. Since that time society has become dependent on crude oil.

Mr. Jacobs tells us that the oil well pumps oil from a reservoir underground. "Does anyone know what an oil reservoir is?" he asks. I imagine it's an underground pool of oil, but Mr. Jacobs says this is wrong. He tells us that *petroleum* means "rock oil." Oil reservoirs are actually areas of rock that have droplets of oil stored in the pores, or spaces, of the rocks. There are no underground pools of oil.

We learn that drilling for oil is expensive. Scientists use sound waves to determine the best places to drill for oil. Sound travels through different types of rocks at different speeds. Computers help measure how fast sound waves move through rock.

"Once we have found an oil reservoir," Mr. Jacobs says, "we drill a production well. As soon as the well hits the reservoir, oil often rises. This is because pressure in the reservoir, from tons of rock lying on the oil, is released. As the pressure in an oil well decreases, the oil will no longer rise on its own. Mr. Jacobs tells us that they install pumps at this point to pump the oil out of the ground."

I look at the many pumps around us. I wonder if all oil is pumped on land.

NORTH DAKOTA OIL BOOM

In 2006, North Dakota began experiencing an oil boom. In 2000, North Dakota produced 89,000 barrels of oil a day. By 2011, production had risen to more than 416,000 barrels a day. This increased production helped the United States export more petroleum products than it imported in 2011, the first time that had happened since 1949. However, the United States still imported large amounts of crude oil. Could improved drilling technologies help the United States and the rest of North America become energy independent, or use energy only from sources in North America?

CHAPTER 3

Offshore Drilling

Not all oil is located underneath land. Much of it is located underneath the world's oceans. To get at this oil, companies need to drill deep into the sea floor. We visit our next destination by helicopter. We land on an offshore oil rig in the Gulf of Mexico. When we get off the helicopter, a petroleum engineer named Christopher Costello meets us. He explains that there are several types of oil rigs, and we are on a semi-submersible rig. These rigs are used in water deeper than 300 feet (90 m).

Drill ships are the next offshore drilling rigs Mr. Costello describes to us. He tells us that these rigs are mounted on ships. A drill pipe can be lowered through an opening on the ship's bottom. They can drill in waters up to 10,000 feet (3,000 m) deep, so they are useful far from shore. Many of these ships can maintain their position over the well using computerized control systems. Otherwise, the ships can be anchored to the ocean floor.

We get much of our oil from oil platforms built to retrieve oil from deep below the ocean's surface.

After exploratory drilling has shown that a large oil reserve exists, the drilling rig is replaced by a more permanent production platform. These platforms are built to last for decades. Workers even have living quarters on the platforms.

Mr. Costello tells us that offshore drilling is a very controversial topic. People who support offshore drilling argue that it will reduce our dependence on oil from other countries and keep gasoline prices lower. There is a lot of oil right off the coasts of the United States, and drilling and selling it within the country would be good for the U.S. economy. Other people believe that offshore drilling will ultimately be bad for the country. Continuing to use gasoline will mean increased carbon dioxide emissions into the atmosphere. These people also worry about the environmental impact of offshore drilling. Accidents are rare, but they do happen. And plants and animals suffer when there is an oil spill. These spills can also affect the livelihoods of people who rely on the environment to earn money. If many fish are killed in an oil spill, people in the seafood business will lose money. If oil washes up on beaches making it unsafe to swim, the hotels and restaurants along those beaches may lose tourists and money as well.

DEEPWATER HORIZON EXPLOSION

On April 20, 2010, the *Deepwater Horizon* offshore oil rig exploded, killing 11 men working on the platform. The rig was drilling an exploratory well in the Gulf of Mexico, about 40 miles (65 km) southeast of the Louisiana coast. The rig burned for 36 hours until it sank into the Gulf on April 22. The 5,000-foot (1,500-meter) riser pipe connecting the well to the platform broke, causing oil to flow into the Gulf of Mexico. Oil spilled for almost three months until the well was capped on July 12, stopping the flow of oil. About 4.9 million barrels of oil were released into the Gulf, causing billions of dollars in damage.

Oil spills can cause great damage to the environment and the economy.

Wow—I can see why offshore drilling is controversial! We thank Mr. Costello for teaching us about offshore drilling. On to our next location!

CHAPTER 4

Turning Oil into Energy

Now that we know how oil is extracted from deep within the earth, my field team wants to know more about how this oil is refined into the fuel that powers our cars and buses. Today my team is visiting an oil refinery in Mississippi. The refinery is gigantic. There are pipes, tanks, towers, and stacks as far as we can see. I notice something else about the refinery—it smells a little like rotten eggs. "What's that smell?" I ask Rosa Martinez, the refinery's senior chemical engineer, who has offered to show us around.

"Crude oil has a lot of different chemicals," she explains. "These need to be removed before the oil can become something useful, like gasoline. The odor you're noticing is the smell of these chemicals being released into the atmosphere. In small amounts, these chemicals won't harm you," she says. "But most refineries are built away from towns and cities for this reason.

It takes a special process to convert oil from the ground into the gasoline that powers our cars.

Still, refineries have to be careful. If a leak or a spill occurs, some of these chemicals can get into the water and be potentially harmful to people and wildlife."

"Crude oil is not very useful when it first comes out of the ground," Ms. Martinez says. "It is made of many different

hydrocarbons. These are chains of carbon and hydrogen atoms. The purpose of the refinery is to separate crude oil into its different parts. Each part forms a different product and has different properties. This is because different hydrocarbons have different numbers of carbons and different arrangements of their chains."

Ms. Martinez explains that the first process occurring at the refinery is this separation. Each of the hydrocarbons

Refining crude oil requires a lot of space and equipment.

present in the crude oil turns from liquid to gas at a different temperature. This allows separation of the parts making up the crude oil. Usually, Ms. Martinez says, the more carbon atoms present, the higher the boiling point. Through a process called fractional distillation, the parts of the hydrocarbons are separated by their boiling point. The mixed hydrocarbons are heated until they turn into gases. The vapors rise through a long column, cooling as they rise higher. As the different parts cool below their boiling point, they condense. This means they change from gas back to liquid.

The final refining process is treatment. After each of the fractions has been separated from the crude oil, refineries must remove the impurities from them. These impurities include substances such as sulfur, nitrogen, oxygen, water, dissolved metals, and salt. Next the different fractions are mixed to create the desired products.

From the refinery, pipelines are usually used to carry the finished products to storage areas. Trucks load the products from there for delivery to customers at locations such as gas stations.

ELECTRIC CARS

Hybrid vehicles use two power sources, usually gas and electricity. Both hybrid and all-electric vehicles help us use oil and coal more efficiently and in ways that are better for the environment. Electric vehicles can convert about 60 percent of the electrical energy to driving power. Gasoline-powered vehicles only convert about 20 percent of the energy stored in the gasoline to driving power. All-electric vehicles don't give off pollution; however, the power plant that generated the electricity usually gives off carbon emissions.

"The refinement process has improved a lot," Ms. Martinez says. "A hundred years ago, only a small part of the oil could be used for transportation fuel. Today, almost the entire barrel of oil can be used. In the past 50 years, improvements in both fuel and vehicles have allowed us to travel about twice as far on a gallon of

Electric-powered cars use gasoline more efficiently than gasoline-only cars.

fuel as we could in 1960." She adds, "Did you know that oil even helps bring food to your table? The trucks that transport food crops across the country run on fuels produced from oil."

I guess oil has its advantages and disadvantages, like any energy source. I thank Ms. Martinez for showing us around. We're ready for our next destination!

CHAPTER 5

Into the Seam

After spending time studying oil, my team has switched gears to focus on another fossil fuel: coal. We are visiting an underground coal mine in Pennsylvania. When we arrive, the mine superintendent, Don Williams, takes us to an elevator. As we ride several hundred feet down a shaft, deep into the ground, Mr. Williams gives us a bit of background on coal. He tells us that like oil, coal is a fossil fuel. It formed from plants that lived in swampy areas more than 300 million years ago. As the plants died, they were covered with more plants, soil, and water. The plants turned to peat, an organic sediment containing partially decayed stems, roots, bark, and twigs. Over time, the pressure from the upper layers squeezed water from the peat. Heat from within the earth caused the peat to break down. Gases were forced out. The peat changed to coal. This process is called coalification.

Like oil, coal is made from ancient organisms compressed over millions of years.

"Coal has been around a long time," Mr. Williams explains. "For thousands of years humans have been using it as a fuel source for heating their homes. During the Industrial Revolution in the 1800s, huge improvements were made in technology in England and other parts of the world. Much of this technology relied on coal as a fuel, increasing coal's role as an energy source. During this time, steamboats and locomotives became important forms of transportation. Their boilers were fueled by coal."

When the elevator finally stops, Mr. Williams tells us we have reached the coal seam. The seam is the layer of coal between two layers of rock. We see a large machine cutting coal from the seam. Mr. Williams tells us this machine is called a continuous miner. On the front of the continuous miner is

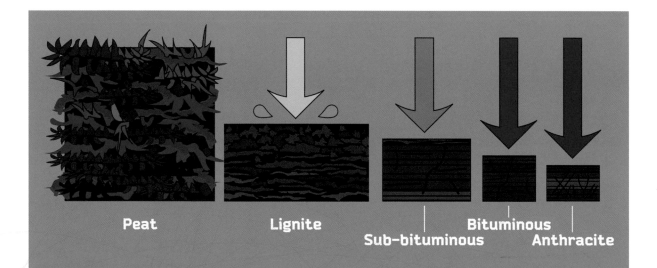

Peat Lignite Sub-bituminous Bituminous Anthracite

TYPES OF COAL

There are four different types, or ranks, of coal. They are lignite, sub-bituminous, bituminous, and anthracite. The different ranks contain different amounts of the element carbon and produce different amounts of heat energy. Lignite is the youngest and lowest ranked coal. It is lighter brown and soft. It has the lowest carbon content and energy potential. Anthracite is the oldest, highest rank of coal. It is dark, hard, and shiny. It has the highest carbon content and energy potential.

Machinery, such as the continuous miner, helps us retrieve coal in a safer way.

a drum covered with rows of teeth. The drum rotates quickly,
allowing the teeth to scrape coal from the seam. The continuous
miner scoops the mined coal and moves it to the back of the
machine. From there, it flows into shuttle cars that carry it
to conveyor belts. The conveyor belts move the coal up to the
surface of the mine where the coal will be processed.

Mr. Williams points to the columns standing throughout the underground mine. "These are pillars. In this mine, we use a mining process known as room and pillar mining. The continuous miner cuts rooms into the coal deposit, leaving the pillars as supports. The pillars keep the roof of the mine from collapsing on us."

DANGEROUS MINING

Coal mines are built with special safety precautions, but mining can still be dangerous. In the past, mines weren't nearly as safe as they are today. Occasionally the mines caved in, sometimes killing everyone inside. Even though mines are much safer today, mining accidents still happen occasionally. Buildups of gas below ground can lead to explosions. In 2010, a coal mine explosion in West Virginia left 29 miners dead.

We ride back up in the elevator. Once we reach the surface, Mr. Williams suggests we look at how the coal is processed and transported. We walk over to the processing plant. Here coal is crushed, sorted, and washed. This removes a lot of the rock and dirt that can't be burned as fuel. We hear a train pulling up to the loading facility and walk over in time to see the train pull its first railcar under the hopper, the temporary storage for the coal. Almost 100 empty railcars are waiting to be loaded. Mr. Williams tells us this trainload of coal is being shipped to a utilities company. The coal will be burned to generate electricity at a power plant. Power plants use a lot of coal; bigger plants

Huge trains transport coal from the mine to a power plant.

may need an entire trainload of coal almost every day. That's a lot of coal! We decide to visit a power plant next to see exactly how a plant turns coal into energy.

CHAPTER 6

Coal Energy

For our next stop, my team visits a coal-fired power plant in Florida to find out how coal is used to make electricity. As soon as we arrive, Stephanie Jackson, the plant manager, meets us. She hands me a small vial containing the coal that the plant uses. This coal isn't shaped like rocks as we had seen at the coal mine. Instead it has been crushed and ground into a fine black powder.

"A power plant has a lot of moving parts," Ms. Jackson tells us. "We feed this coal into a firebox where it mixes with air and burns. The firebox provides heat to a big boiler at the power plant. We pump water through pipes inside the boiler. As the coal burns in the firebox, it heats the water inside these pipes. Once the water becomes hot enough to boil, it changes into steam. The steam is piped to a turbine, a machine with large blades. The pressure of the moving steam turns the blades located in the turbine. As these blades turn, they spin the turbine shaft."

Coal-fired power plants provide about half of the electricity used in the United States.

Ms. Jackson tells us that the turbine shaft is connected to a generator. As the shaft turns, the generator changes the spinning mechanical energy into electricity. This is done through magnets that spin in the generator. The spinning magnets produce an electric current in wire coils.

After completing its work in the turbine, the steam is sent to the basement of the power plant into a condenser. The condenser is a large chamber with tubes of cool water running through it. The steam condenses on the tubes, changing back into water. The water can then be reused in the boiler. "Large amounts of water are required for this cooling process," Ms. Jackson adds, "which is why this power plant is next to a river."

THOMAS EDISON'S PEARL STREET GENERATING STATION

Inventor Thomas Edison created a working light bulb in 1879. But in order to use it and sell his idea, he needed a way to deliver electricity to customers. In 1882, Edison oversaw the installation of electric conduits and wiring underneath some of New York City's streets. Six giant dynamos were installed on Pearl Street to light up one square mile (2.6 sq km) of the city. The dynamos, known today as generators, produced electricity. Coal was fed into the plant's boilers to provide steam for the dynamos. Pearl Street Station went into service in 1882 and was one of the earliest power-generating plants. Technology, however, changed. Edison's plant could not meet the newer technology's needs. The plant was dismantled in 1895.

After the steam condenses on the tubes in the condenser, the water inside the tubes is hotter. This hot water is then sent to the cooling towers. Ms. Jackson also explains why we sometimes see white clouds above the cooling tower. She says the low-temperature steam entering the cooling tower comes into contact with cool air, cooling the steam to form water droplets in the air. These water droplets may create the clouds we sometimes see. She says this is the low-temperature steam used inside the turbine. It is not smoke, as many people think.

"Coal-fired power plants are more efficient than many other energy sources, such as solar power, and are relatively inexpensive to operate. Right now, about half of the country's electricity needs are met through coal power," Ms. Jackson tells us. "Plus, even though coal is a non-renewable source, we have a lot of it and much of it can be mined and sold right here in the United States, which is good for our economy."

She adds, "But as with all energy sources, there are some downsides." For one, Ms. Jackson explains, coal-fired power plants release a lot of carbon dioxide gas. This carbon dioxide can be bad for the environment.

"Even though coal power is more economical than other energy sources, most older coal plants are only about 30 percent efficient," Ms. Jackson says. "Think about it this way—if you were cleaning your room at 100 percent efficiency, it would take you an hour. If you were cleaning it at 1 percent efficiency, it would take you 100 hours. New technology can help make plants more efficient," Ms. Jackson adds. "But adapting older plants to the new technology will be expensive."

Thomas Edison was a pioneer in using coal power to create electricity.

Our trip to the power plant has given my team a lot to think about. Ms. Jackson said there is technology that can make a coal-fired power plant more environmentally friendly. We decide to investigate further.

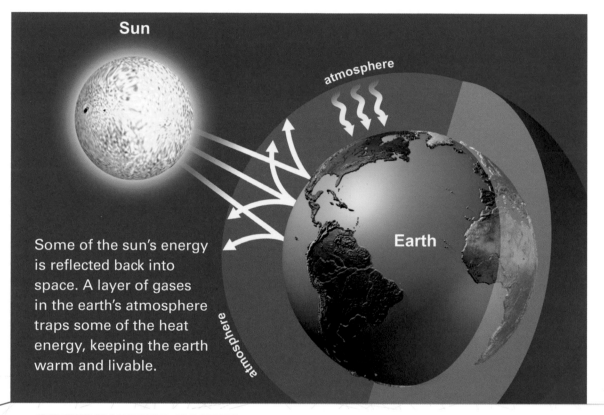

Some of the sun's energy is reflected back into space. A layer of gases in the earth's atmosphere traps some of the heat energy, keeping the earth warm and livable.

GREENHOUSE GASES

When the sun heats the earth, the earth absorbs some of the sun's energy while other energy is reflected to space. A natural layer of gases in the atmosphere traps some of this heat, warming the earth. This is called the greenhouse effect. The gases in the atmosphere are called greenhouse gases. Without the greenhouse effect the earth would be much colder and covered in ice. However, humans are adding greenhouse gases to the atmosphere. Many scientists believe these added gases cause more heat to be trapped on the earth, increasing our planet's average temperature.

CHAPTER 7

Cleaner Coal

After learning about some of the pros and cons of coal power at the Florida plant, our field team wanted to learn more about the ways to control the air pollution created by burning coal in power plants.

Today, we head to a coal-fired power plant in Ohio that has recently constructed a new scrubber, a device that removes pollutants from the gases a coal-fired power plant creates. When we arrive, Greta Weinstein, the power plant's environmental engineer, greets us. She tells us we can call her Greta and invites us to join a training session at the plant to learn how the new scrubber works.

We learn that although carbon dioxide is one of the downsides to coal power, it is not the only pollutant released when coal is turned into energy. Sulfur is a yellow substance found naturally in coal. Some of this sulfur was in the ancient plant life from which the coal formed. Other sulfur came from ancient seawater that had contact with the coal as it

A scrubber is a tall tower that helps a coal-fired power plant control air pollution.

was forming. As coal burns, the sulfur combines with oxygen and forms sulfur dioxide, a poisonous gas that can lead to acid rain.

Federal clean air regulations limit the amount of sulfur dioxide power plants can release when they burn coal. Because of this, all power plants built after 1978 must include a sulfur-removal system in their coal-conversion process if they are burning a high-sulfur coal. One way to remove sulfur is by simply crushing the coal and washing it before burning it. Impurities, such as sulfur compounds, will sink to the bottom and coal will float during washing.

After combustion, the stage where the coal is burned, the smoke created passes through the scrubber. The plant's cylindrical, tower-shaped scrubber removes sulfur from the power plant's exhaust gases before it is released into the air. Greta says that the process of removing sulfur in the scrubber is called flue gas desulfurization.

I raise my hand. "Is sulfur the only substance that causes air pollution from a power plant?" I ask.

"No, it isn't," Greta answers. "We also remove nitrogen oxides at our power plant."

She explains that nitrogen makes up about 78 percent of the air we breathe. At the high

The sulfur naturally found in coal releases harmful gases when it is burned.

ACID RAIN

Burning fossil fuels can release sulfur dioxide and nitrogen oxides into the atmosphere. When these gases react with water, oxygen, and other chemicals in the atmosphere, mild solutions of sulfuric acid and nitric acid are formed. When these acids fall to Earth as precipitation, in the form of rain, snow, fog, or even dry particles, it is called acidic deposition or, more commonly, acid rain. Acid rain causes lakes and streams to become acidic, a process called acidification. This can harm aquatic life. Acid rain also damages trees. It destroys paints, buildings, statues, monuments, and bridges. Government rules and more awareness of the issue have reduced the amount of acid rain that falls today, but it is still a problem.

temperatures at which coal is burned, nitrogen in air can combine with oxygen to form nitrogen oxides. Nitrogen oxides contribute to respiratory problems as well as to the formation of smog, a brown haze that hinders visibility.

Greta says that one method for reducing nitrogen oxide emissions uses low-nitrogen oxide burners, which prevent nitrogen oxide from forming in the first place. These burners reduce the amount of oxygen going to the primary combustion area of the burner. With reduced oxygen, the coal does not burn completely. It is then sent to another chamber to repeat a similar process. This occurs until all coal is burned up.

Still, she says, it's impossible to make coal completely pollution-free. The scrubbers help remove most of the pollutants, but they can't get them all. Small amounts of pollutants still make their way into the environment.

CHAPTER 8

A Bright Future?

Our last trip is to a state-of-the-art coal gasification and carbon capture research center in Alabama. The center's mechanical engineer, Mohammed Assan, meets us as we arrive. He tells us that the research center is conducting a pilot program using two new technologies. The first is coal gasification. The second is carbon capture.

Mr. Assan tells us that coal gasification is one of the cleanest methods for converting coal into electricity. "The key to coal gasification is a vessel called a gasifier," he explains. "Rather than burning the coal, a gasifier breaks the coal apart using a combination of heat, pressure, and steam. This process produces synthesis gas, known as syngas. Syngas is primarily made up of hydrogen and carbon monoxide and can be burned."

Mr. Assan tells us that next the syngas is cleaned of impurities such as sulfur and ammonia. Then, it is burned in a two-step power generation process. First, the syngas is burned in the combustion turbine. The heat from the combustion

Engineers and scientists are working hard to design and build new power plants that are more efficient and better for the environment.

turns a generator to produce electricity. Next, the leftover heat from the combustion turbine is used to boil water. This creates steam to turn a steam turbine, also producing electricity. The use of these two types of turbines is called a combined cycle. "Coal gasification power plants have high power generation efficiencies," Mr. Assan explains. "They also produce less carbon dioxide than traditional coal-fired power plants, so they are cleaner." Mr. Assan tells us gasification has been around for a long time. But as our coal supply keeps getting lower, it will be more important to develop and refine new technologies to make coal use more efficient and make our supply last longer. Right now, gasification plants operate at about 50 percent efficiency. Future gasification plants might approach 80 percent efficiency.

The second technology that Mr. Assan explains to us is carbon capture. He says most power plants burn fossil fuels. This releases large amounts of carbon dioxide into the atmosphere. Carbon capture is a new technology that traps the carbon dioxide from combustion before it is released into the atmosphere. After the carbon dioxide is captured, it is piped deep underground. There it is permanently stored in rock formations, often more than a half mile (.8 km) underground. The site is monitored to make sure the carbon dioxide doesn't escape back

RECLAIMING COAL MINES

Surface mining of coal disturbs the land. But modern mining practices require mines to be able to be roused, or reclaimed, after the mine is closed. Reclamation restores the land, returning it to a useful purpose. Reclamation involves replacing topsoil that was stripped away. Slopes are evened out. The mining area often becomes a lake or pond. Grass is usually replaced by hydroseeding, spraying a mixture of seed and fertilizer over the ground. Trees are replanted. Reclaiming coal mines also makes the land more attractive to wildlife. Some reclaimed mines can be used as golf courses, fishing areas, or parks.

Most power plants use a steam-powered turbine like this one to create electricity. Gasification adds a second turbine, making the power plants more efficient.

to the surface. Mr. Assan said that plants using carbon capture also need to make sure carbon dioxide doesn't get into the groundwater people use for drinking.

"Right now, commercial carbon capture plants are very uncommon," he explains. "For one thing, the technology uses a lot of energy and is expensive. If plants were to start using it full time, they would have to raise their prices to pay for it. Your parents' electric bill would probably go up. Still," he adds, "technology will keep improving. New coal plants being built have much stricter regulations than old plants."

If this is true, coal energy will keep getting cleaner. Our team has learned a lot from Mr. Assan. Like oil, coal is an important part of our economy and meeting our energy needs. It is important for us to protect this resource and to make the most of it, while also protecting our environment by finding ways to use coal energy responsibly.

You can help our coal supply last longer by conserving energy. Try replacing your incandescent lightbulbs with LED lightbulbs, which use less energy to light your house.

CHAPTER 9

Your Turn

You've had a chance to follow Jamaal and his field team as they conducted research on oil and coal. Now it's time to think about what you learned. Coal and oil play a huge role in meeting our current energy needs. They are both fossil fuels that need to be removed from the ground. They can be burned to create heat. This heat can be used directly, such as in a furnace. It can power a turbine to produce electricity. It can also provide transportation energy. Right now, coal and oil are readily available. They will be key energy sources for many years to come.

Coal and oil are also non-renewable resources that will eventually run out. Both create greenhouse gases and other pollutants that can harm the environment. But new methods and technology can help keep these fossil fuels more environmentally friendly and more efficient. More efficient fossil fuels mean our coal and oil supply will last longer. We need to do our best to conserve these non-renewable resources.

YOU DECIDE

1. Do you think the pros of coal and oil power outweigh the cons? Why or why not?

2. Take a look around your house and garage. How many of the things you see are powered by coal and oil?

3. Think about the advancements being made in coal and oil energy. Which technology discussed has the brighter future? Why?

4. There are positive and negative aspects to all methods of retrieving oil and coal from the earth. Which method do you think io the most positive? Which is the most harmful? Why?

5. What can you do to cut down on your energy use? Think about technology and ways to change your behavior.

What do you think? Should we try to find ways to use our coal and oil more efficiently? Or should we focus on finding alternative energy sources? Or should we do both?

GLOSSARY

acid rain: The result of sulfur dioxide and nitrogen oxides reacting in the atmosphere with water and returning to Earth as rain, fog, or snow.

barrel: The 42-gallon unit used to measure the volume of crude oil.

coal seam: The layer of carbon-rich material located between layers of rock.

coalification: The process that converts plant material into coal due to heat and pressure in the earth.

combustion: A chemical process where something combines with oxygen to burn.

crude oil: Unrefined petroleum (oil).

fossil fuel: A fuel formed in the earth over millions of years from plant and animal remains, including coal, oil, and natural gas deposits.

generator: A machine that converts mechanical energy into electricity.

greenhouse gas: Gas in the atmosphere, such as carbon dioxide or water vapor, which absorbs heat radiated from the earth's surface.

hydrocarbons: Organic chemical compounds composed of carbon and hydrogen.

organic: Material that originated from plants or animals.

petroleum: Thick yellowish black liquid mixture of hydrocarbons.

pores: Open spaces in rocks that can contain air, water, or petroleum.

power plant: A factory that generates electricity.

refinery: A factory that processes and purifies crude oil.

scrubber: A device that removes impurities or pollutants from a gas.

turbine: A machine that uses a fluid to produce a spinning action. Turbines spin electric generators.

EXPLORE FURTHER

Be a Conservationist

Fossil fuels formed over millions of years. They are a non-renewable resource. Once they're gone, they're gone forever. But, you can conserve fossil fuels by reducing the amount of electricity and petroleum products you use. Turn off lights and televisions when you are not using them. Don't leave doors open for air conditioning or heat to escape. Even recycling aluminum cans saves energy. Look around your home and make a list of ways you can help conserve fossil fuels. Follow this list for a week, taking notes on which of these ways are easy to adapt to and which are hard.

Oil Emergencies

The U.S. Strategic Petroleum Reserve is the world's largest emergency stockpile of oil. The president can issue a drawdown, or release, of oil from the reserve when an emergency threatens our national supplies. Visit the Department of Energy's web site and download their study guide on oil to find out what circumstances have resulted in emergency drawdowns of the reserve.

Reduce Your Carbon Footprint

The amount of greenhouse gases you produce is sometimes called your carbon footprint. Visit an online carbon footprint calculator to estimate how much carbon dioxide your household produces in a year. Examine your results—where can you reduce emissions? Can you turn off the lights every time you leave a room? Can you replace outdoor electrical lighting with solar-powered lamps? What about growing your own food to reduce driving trips to the grocery store? What are other things you could do to reduce emissions?

How can you help conserve energy?

SELECTED BIBLIOGRAPHY

"Fossil Energy Study Guide: Oil." *U.S. Department of Energy*, n.d. Web. Accessed May 21, 2012.

"Fossil Energy: How Gasification Power Plants Work." *U.S. Department of Energy*, n.d. Web. Accessed June 9, 2012.

"The Coal Resource: A Comprehensive Overview of Coal. *World Coal Institute*, March 6, 2009. Web. Accessed May 22, 2012.

Stultz, S. C., and John B. Kitto. *Steam, Its Generation and Use, 40th Edition*. Barberton, Ohio: Babcock & Wilcox, 1992. Print.

FURTHER INFORMATION

Books

Aaseng, Nathan. *Business Builders in Oil*. Minneapolis: The
Oliver Press, Inc., 2000.

Horn, Geoffrey M, and Debra Voege. *Coal, Oil, and Natural Gas*.
New York: Chelsea Clubhouse, 2010.

Morgan, Sally. *The Pros and Cons of Coal, Gas, and Oil*. New York:
Rosen Central, 2008.

Solway, Andrew. *Fossil Fuels*. Pleasantville, NY: Gareth
Stevens, 2008.

Websites

http://www.adventuresinenergy.org
Research the processes involved in getting oil and natural gas to
the consumer.

http://www.eia.gov/kids
Follow the energy links for information on different energy
sources and their histories, including energy-related games
and activities.

http://energyquest.ca.gov/story/index.html
Follow *The Energy Story* to find out all about the energy that
makes the world work.

INDEX

acid rain, 32, 34

barrel, 9, 12, 18

carbon, 5, 12, 16–17, 18, 22
carbon capture, 35, 37–38
carbon dioxide, 12, 29, 31, 36–38
coal gasification, 35–36
coal mines, 20, 24
coal seam, 22–23
coal-conversion process, 32–33
coalification, 20
combustion, 33–34, 35–37
condenser, 27–28

Drake, Edwin L., 7, 8
drilling, 7, 9, 10–13

Edison, Thomas, 28, 29
efficiency, 6, 18, 28–29, 36–37, 40
electric cars, 18, 19

fractional distillation, 17

greenhouse gases, 30, 40

hybrid vehicles, 18
hydrocarbons, 16–17

impurities, 18, 33, 35
Industrial Revolution, 21

natural gas, 5, 7
nitrogen, 18, 33–34
North Dakota, 9

offshore drilling, 10–13
oil refineries, 14, 16, 18
oil reservoirs 8–9
oil spills, 12, 15

Pearl Street generating station, 28
peat, 20
pillars, 24
pollution, 18, 31–34, 40
power plants, 18, 24–25, 26–30, 31–33, 36–39
pumps, 8–9

scrubber, 31, 33, 34
semi-submersible rigs, 10
steam, 26–28, 35–36
sulfur, 18, 31–33, 34, 35
syngas, 35

turbines, 26–28, 35–36, 40